Be An Expert!™

Ocean Animals

Amy Edgar

Children's Press®
An imprint of Scholastic Inc.

Contents

Know the Names

Be an expert! Get to know the names of these ocean animals.

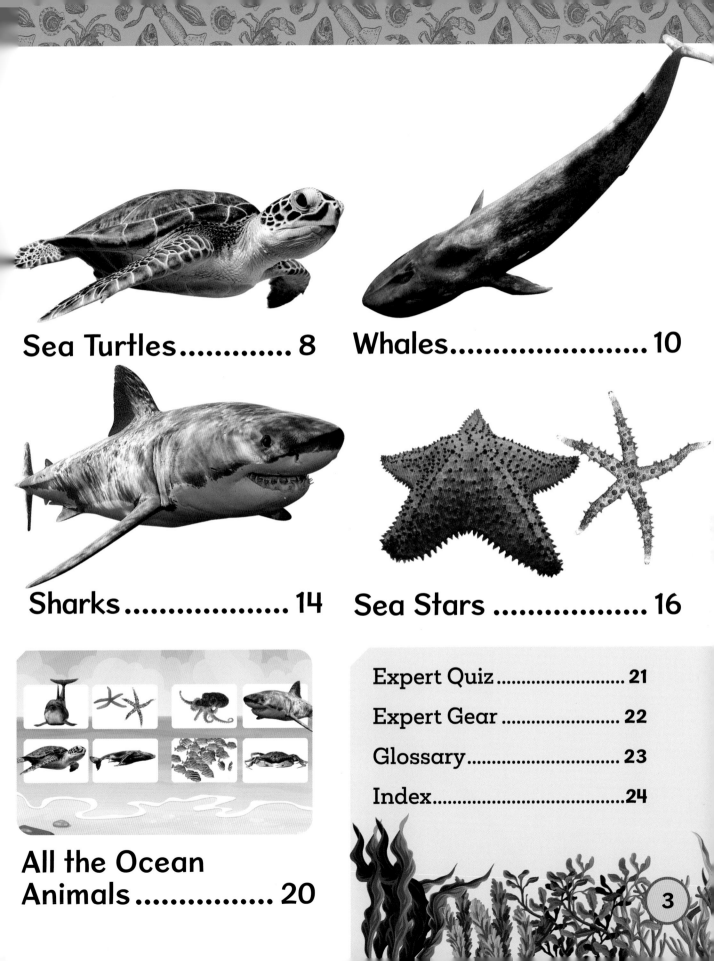

Dolphins

They are playful.
They **leap** and splash!

Deep Dive

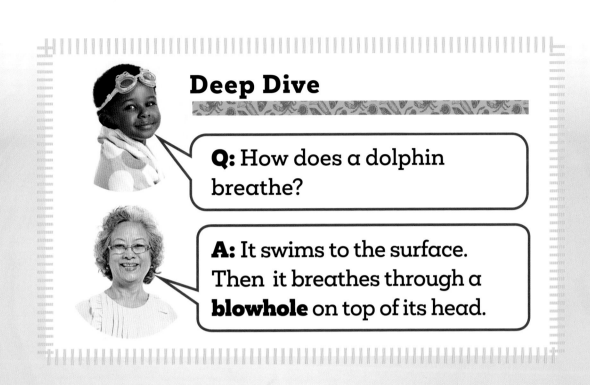

Q: How does a dolphin breathe?

A: It swims to the surface. Then it breathes through a **blowhole** on top of its head.

blowhole

Crabs

They have 10 legs.
Most walk sideways.

king crab

snow crabs

Crabs shed their shells as they grow bigger. Then they grow new ones. This process is called molting.

blue crab

Dungeness crabs

Sea Turtles

They are great at swimming and diving. Their shells protect them.

Zoom In

Find these parts in the big picture.

beak   **shell**   **claw**   **flippers**

Whales

They are the biggest animals on Earth. Some are almost as long as two school buses.

blue whale

Expert Fact

A whale's tail has two **flukes**. The flukes move up and down to push the whale through the water.

humpback whale

Octopuses

They have eight strong arms.

Deep Dive

Q: How do octopuses protect themselves from **predators**?

A: They change color or squirt ink.

Sharks

Chomp! Sharks have lots of teeth.

great
white shark

tiger shark

Zoom In

Find these parts in the big picture.

eye

pectoral fin

tail

gills

saw shark

hammerhead sharks

Sea Stars

They have spiny arms.
If they lose an arm, they
grow a new one.

Deep Dive

Q: What do sea stars eat?

A: Sea stars eat clams and oysters. They have tiny suction cups under their arms. They use them to pry open the shells.

Fish

They breathe underwater.

stripey

yellowfin
goatfish

cod

blue jack mackerel

tuna

Expert Fact

Most fish do not have eyelids. Their eyes are always open, even when they sleep.

All the Ocean Animals

They are amazing swimmers.
Thanks, ocean animals!

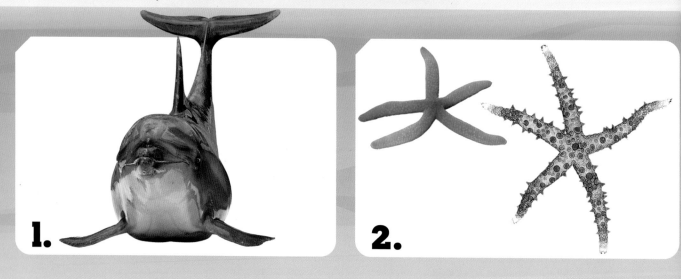

1.

2.

5.

6.